Alien Pi in the Sky?

Also by same author

Matrix Magic ©2010

A Little Rocket Science and other Baubles ©2013

Alien Pi in the Sky?

By
Olatunde Adeyemo

Olatunde Adeyemo
2016

Copyright ©2016 by Olatunde Adeyemo

All rights reserved. No parts of this publication may reproduced, or stored in any retrieval system, or transmitted in any form or means, electronic, mechanical, photocopying, recording or otherwise without the prior written permission of the author except for the use of brief quotations in a book review or scholarly journal.

First Printing: 2016

ISBN 978-1-326-72720-8

www.anglo-african.com

Distributer: http://www.lulu.com

Dedication

To my late father John Oladeji Adesokan Adeyemo, who taught me the importance of principal and integrity.

Contents

Preface xii
Synopsis 1
Chapter 1 Helical Springs 4
Chapter 2 3Dimensional Stess Analysis 13
Chapter 3 Pi in the Sky and Extended Precision Calculation 45
Study of Surds 54
Evaluation of Pi-1 55
Evaluation of Pi-2 64
Bibliography 67

List of Figures

Figure 1.1 torsion 4

Figure 1.2 shear deformation...5

Figure 1.3 Helical Spring..8

Figure 2.1 modes of deformation 14

Figure 2.2 Parallel radial and transverse force components 17

Figure 2.3 Extension, bending and torsion of beam A caused by force F at P 18

Figure 2.4   2nd moment of area 20

Figure 2.5 simple lever 25

Figure 2.6 Input file... 26

Figure 2.7 3DENetOut.txt... 28

Figure 2.8 5 Beam cantilever 30

Figure 2.9 A complex multi loaded structure.............38

Figure 2.10 Tripod Frame 42

Figure 3.1 Bounding internal and external squares 56

Figure 3.2 Geometric Estimation of Pi 60

List of Tables

Table 2.1 Properties of metals 23
Input File 31
node 6 onwards 34
node 3 onwards 36
Original position of nodes 39
with nodes 3,4 and 5 loaded 40

Preface

I would conjecture about aliens.

I would ponder over Pi.

I would peer and interrogate the sky.

Could I fabricate a construct to promote a supposition?

That if true, would fundamentally alter your condition.

I leave you to cogitate over this manual and reconsider your position.

Could there be Alien Pi in the Sky?

Olatunde Adeyemo

Synopsis

In this document we examine theories for the mechanics of helical springs, stress in lattice like structures and methods for calculating Pi (π).

I show how the theory of springs is used to calculate a helical spring's load bearing capacity and show an example of design and improvement.

In "three dimensional stress analysis", I build a model to analyse the stress and strain reactions of members of a structure comprised of beams and nodes, the beams connect the nodes in some form of lattice. The model is then converted into a program that can be worked. We then look at a series of increasingly complex structures and look at the results of them being loaded. Starting from a simple lever end loaded with a single load leading onto structures with multiple loads with multiple limbs connected by nodes with various constraints and more than one single fixed node and finally a full 3 dimensional lattice.

To be able to create an extended precision answer I needed to create an arithmetic system that could cope with doing operations accurately to a specified number of decimal places. This was a serious task, requiring development and refining for it to work well. I use this system to generate surds that have an easy method for result verification. I then go on to show how we can use this system to produce ever more accurate values for Pi using two different methods. We can finally use the Machin formula as an efficient generator of Pi. These methods may be utilised for the testing of computing ability of different computers.

Alien Pi in the Sky?

The programs, code, examples of input and output files and study compilations used in this document are available for use with acknowledgement and can be downloaded from www.Anglo-African.com/AlienPi.

Olatunde Adeyemo

Alien Pi in the Sky?

Chapter 1   Helical Springs

Have you ever considered how a spring coil functions? The coils in your car's suspension that make your ride so comfortable, or the large ones often seen in the undercarriage of trucks and railway wagons. The principle is the same as for bending or compression of an elastic member and works on the basis of Hooke's Law.

In torsion a prismatic member experiences a torque couple causing an angular displacement as shown in fig 1.1

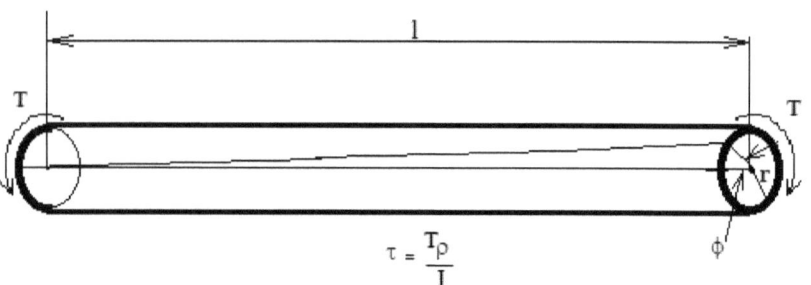

$$\tau = \frac{T\rho}{J}$$

Figure 1.1 torsion

If we consider a short section of a prismatic spar under torsion strain as in fig 1.2, it is evident that the shear strain at radius ρ from the centre is $\rho \, d\phi$. This will be maximum at the outermost surface of the member.

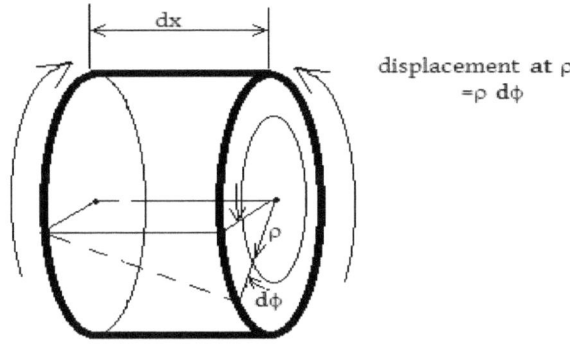

Figure 1.2 Shear deformation

The formula for displacement per unit length is

$$eqn\ 1.1 \quad \gamma = \rho \frac{d\phi}{dx}$$

, but we may use

$$eqn\ 1.2 \quad \theta = \frac{d\phi}{dx}$$

$$eqn\ 1.1b \quad \gamma = \rho\theta$$

We have shear stress

$$eqn\ 1.3 \quad \tau = G\gamma = G\rho\theta$$

Alien Pi in the Sky?

G is shear modulus, described as

eqn 1.3b $\quad G = \dfrac{\tau}{\gamma}$

And the applied torque across the section

eqn 1.4 $\quad T = G\theta \int \rho^2 dA = G\theta J$

J 2nd moment of area

eqn 1.5 $\quad J = \int_A \rho^2 \, dA$

If the member is of circular cross section

$$J = \dfrac{\pi r^4}{2}$$

If the member is circular but hollow with inside and outside radii r_i and r_o

$$J = \dfrac{\pi}{2}(r_o^4 - r_i^4)$$

Torsion properties are responsible for the functioning of helical springs. A spring loaded with load P centrally directed through it as in fig 1.3 essentially applies a moment PR on a member cross section.

There are assumptions that we make to simplify our handling of helical springs. The pitch between successive coils is assumed to be small compared to the radius of the spring and the diameter of the coil cross section is considered small compared to the radius of the spring and any deformations created by loading the spring are small compared to the overall dimensions. These are equivalent to R>>ρ, R>>q and R>>δ.

The rest length of the spring with n turns is
$$l = n\sqrt{(2\pi R)^2 + q^2}$$

The load PR results in a compression such that

$$RP = G\theta \int \rho^2 dA = G\theta J$$

The largest filament extension = $l\theta\rho_m$ with $\rho_m = r$

And a displacement δ

$$\delta = l\theta R$$

eqn 1.6 $\delta = \theta n R \sqrt{(2\pi R)^2 + q^2}$

Eqn 1.3 gives us the maximum stress

$$\tau_m = G\gamma = G\rho_m\theta, \quad \tau_m = Gr\theta$$

Alien Pi in the Sky?

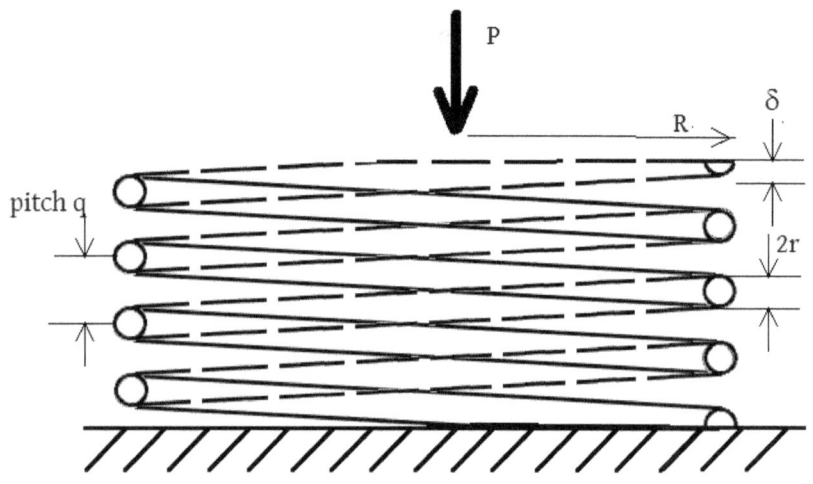

Figure 1.3 Helical Spring

Example 1

A coil spring of 6 turns formed from steel wire of diameter 6 mm with pitch 4 mm, radius 160mm, G=81 * 10^9 Nm^{-2}.

a. Find the load that will cause a 1cm depression of this spring. Find the maximum stress created by this load.

b. Find the depression caused by this load in a spring of tubular section with central void radius of 3mm, the same cross sectional area and with all other dimensions similar.

Soln.

1cm depression means $\delta = 0.01\text{m} = l\theta R$

Therefore $\theta = \delta/(lR)$

where $l = n\sqrt{(2\pi R)^2 + q^2}$ =

$n*\sqrt{4\pi^2 * 0.16^2 + 0.004^2} = 6.032\text{m}$

$\theta = 0.01/(6.032*0.160) = 0.01/0.9651 = 1.04*10^{-2}$ rads/m

$\tau_m = 81*10^9 * 3*10^{-3} * 1.04*10^{-2} = 2.53*10^6$ Nm^{-2}

$\sigma_y = 2.5*10^6$ Nm^{-2}

The ratio of $\tau_m/\sigma_y = 1.01$

This is clearly likely to fail.

To find the load that causes this deflection use eqn 1.4

$PR = G\theta J$

$J = \dfrac{\pi r^4}{2} = \pi*(3*10^{-3})^4/2$

$P = \dfrac{G\theta J}{R} = 81*10^9 * 1.04*10^{-2} * \pi*(3*10^{-3})^4/(2 *0.160)$
$= 0.670\text{N}$

Part b.

The cross sectional area of rod radius 3mm is given πr^2
= $\pi*(3*10^{-3})^2 = 2.827*10^{-5}$ m^2

For the tube section $\pi(r_o^2-r_i^2) = 2.827*10^{-5}$ m^2

$r_o^2 = 2.827*10^{-5}/\pi +(3*10^{-3})^2 = 18*10^{-6}$

$r_0 = 4.24*10^{-3}$

$P = \dfrac{G\theta J}{R}$

$\theta = \dfrac{PR}{GJ}$

$J = \dfrac{\pi}{2}(r_o^4 - r_i^4) = \dfrac{\pi}{2}((4.24*10^{-3})^4 - (3.0*10^{-3})^4)$

=3.80*10^{-10}

θ =0.67*0.16/(81*10^9 *3.80*10^{-10}) = 3.48*10^{-3}

The max stress is now given

$\tau_m = G\rho_m\theta$

τ_m =81*10^9*4.24*10^{-3}*3.48*10^{-3} = 1.20*10^6 Nm^{-2}

τ_m/σ_y =0.478

This may be sufficiently within tolerance to be acceptable.

Thus it demonstrates how the same amount of material can be used to make a more effective spring, that may take a higher load or reduce the surface stress, In other words how the design of a spring may be optimised by placing the load bearing material at larger radius.

Olatunde Adeyemo

A fuller and more detailed analysis of helical springs is available on the seminal and definitive work "Mechanical Springs" by A M Wahl 1944. In this work he elaborates on ways of handling springs that do not necessarily fit the assumptions we have made.

Chapter 2 3 Dimensional Stress Analysis.

In this chapter we would like to examine the flexing of 3 dimensional structures under load.

In my previous volume "A Little Rocket Science and Other Baubles ", in chapter 2 "A Heuristic Approach to Solution of Equilibrium In Statically Indeterminate Structures" I was able to demonstrate the use of network loops to solve for equilibrium solutions of 2 dimensional lattices of elastic members. I also made assertion that the same approach could also be applied to a full 3 dimensional structure. In this chapter I look further into it to demonstrate the use of this method.

Given a simple long prismatic member with simple small elastic deformations, the deformation comprises three elements. The first being compression or tension along the length causing a linear lengthening or shortening of the member, the most elemental type of deformation.

Alien Pi in the Sky?

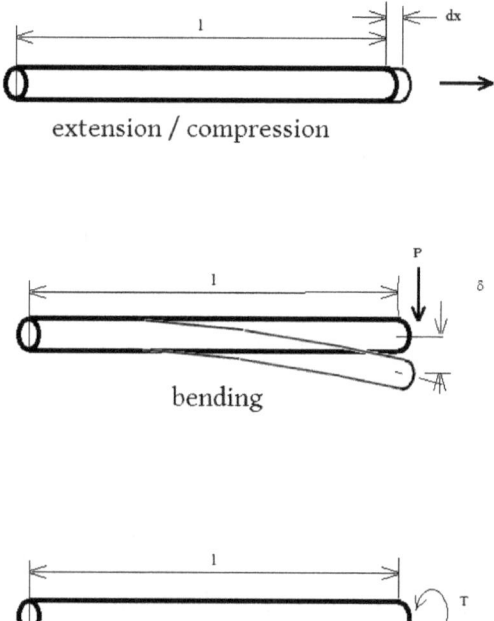

Figure 2.1 modes of deformation

The second type of deformation of this member occurs if the forces are coplanar tangential to the line of direction of the member and since the member is not in motion, they form some kind of bending couple that forces the centre of the member to be out of line with the ends.

Finally the member may experience a twisting couple that creates torsion forces within the member.

In my previous work " A Little Rocket Science and Other Baubles", 2013 the chapter "A Heuristic Approach to Solution of Equilibrium in Statically Indeterminate Structures" resolves planar forces on static spars into tensional or compressive forces that affect the length of spars and equilibrates forces across the structure, a model limited to using only elastic extension or compression of spars and not including bending or torsion was used, for the sake of demonstrating the principles of the method. In this chapter I intend to show the full 3dimensional resolution of forces.

As before, we must define our network in turns of nodes and connectors. For this sake we consider our structure to comprise of beams joined by nodes.

The nature of the nodes is that they have 3 distinct attributes. They may or not be able to shift position, ie they may or not have an invariant position in the coordinate system. The nodes that remain static are known as "feet" of the structure. The nodes each have to be considered whether or not they are able to transfer force through bending or whether the angles between spars are allowed to vary. Thirdly each node has to be considered if torsion forces will be allowed to transfer through it or if the spars attached are allowed to rotate freely. At least one of the nodes will have an external force acting on it that it transmits to the structure.

In our structure we have a possibility in each beam to have extensive/contractive forces, bending forces and torsion forces. In real life the actual combination of these forces are decided by force directions and whether forces are applied as couples and the physical nature of force transmission at the nodes of the structure. To be able to continue a general analysis as we are performing here we may need to make a few assumptions and generalizations.

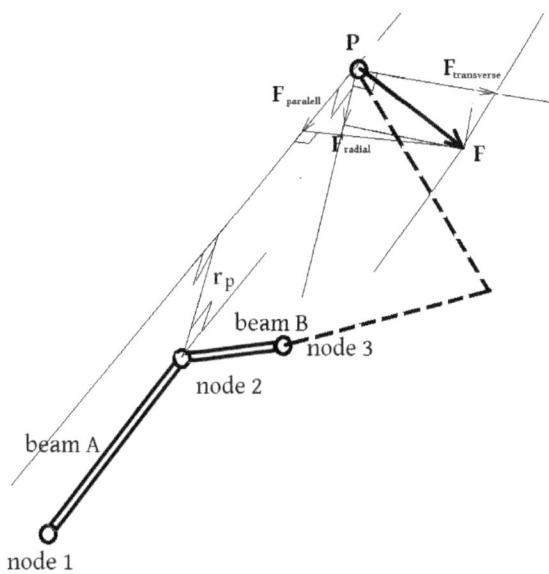

Figure 2.2 Parallel radial and transverse force components

Two beams A and B are joined by node 2 in a planar system represented in figure 2.2. Force F is transferred through some rigid members to node 2 connecting the beams. If everything is rigid the force F applied at P can be considered to transfer in terms of 3 components. The component parallel to beam A produces compression or tensional forces at node 2 through to node 1. This parallel component of force if applied off the central line of beam A will create a moment that is dependent on the radial distance of application away from medial line of beam A, this will cause bending in beam A. The component of force radial that directs from the point of application through the

Alien Pi in the Sky?

medial of beam A will cause a bending moment about node 1 and beam deformation in beam A. The component of force at right angles to both parallel and radial force components will be in tangential to beam A and will contribute to a torsion action on beam A.

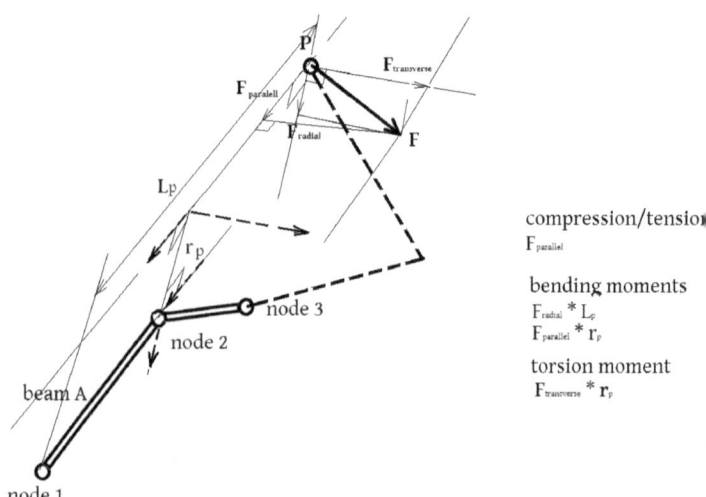

Figure 2.3 Extension, bending and torsion of beam A caused by force F at P

Moment for bending in beam A = $F_{radial} * L_p + F_{parallel} * r_p$

Moment for torsion in beam A = $F_{transverse} * r_p$

In this way we are able to demonstrate how the resolved force acting at P has effect on beam A and consequently estimate a movement in node 2. This in turn works down the limb through each beam up to the point of the free acting force.

How does this work through a complex structure with a number of feet and multiple loads? In This case we must set up the virtual loops in the structure that are to be solved. Each loop comprises a path from a foot through nodes and beams to a free load. Each foot will be responsible for a proportion of the load and each member of the loop will carry this portion of load with the subsequent deformation associated. For each member the cumulative effect of each loop in which it is associated is additive as also are the deformations.

Deformations are non linear so to keep distortions to a minimum we assume they are small and limit our conditions to keep them so.

In working out the deformation due to bending we use the formula

$$Px = \frac{EI_y}{r}$$

As shown in figure 2.1 of "A Little Rocket Science and other Baubles".

If we have circular cross section we can show as in fig 2.4 below the elemental second moment of area

$= (r\sin\theta)^2(r\sin(\theta+\delta\theta)-r\sin\theta)r\cos\theta$

Alien Pi in the Sky?

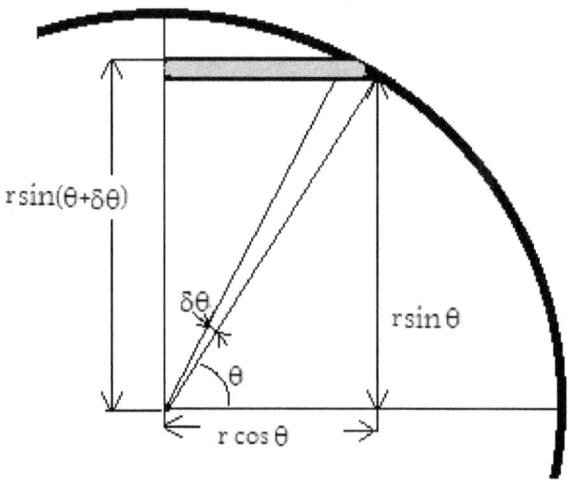

2nd moment of area of greyed part=
$(r\sin\theta)^2 (r\cos\theta)(r\sin(\theta+\delta\theta) - r\sin\theta)$

Figure 2.4 2nd moment of area

We may use the trig identity
$\sin(\theta \pm \phi) = \sin\theta\cos\phi \pm \sin\phi\cos\theta$

To show
$\sin(\theta + \delta\theta) = sin\theta cos\delta\theta + cos\theta sin\delta\theta$

$\delta\theta \ll 1$, so

$$\sin\delta\theta \approx \delta\theta - \frac{\delta\theta^3}{3!}$$

$$\cos\delta\theta \approx 1 - \frac{\delta\theta^2}{2}$$

$$\sin(\theta + \delta\theta) \approx \sin\theta + \delta\theta\cos\theta$$

$$\sin(\theta + \delta\theta) - \sin\theta \approx \delta\theta\cos\theta$$

The second moment of area is then

$$I = 4r^3 \int_0^{\frac{\pi}{2}} (\sin(\theta + \delta\theta) - \sin\theta) r\cos\theta \sin^2\theta$$

$$= 4r^4 \int_0^{\frac{\pi}{2}} \cos^2\theta \sin^2\theta \, d\theta$$

Using trig identities

$$\cos^2\theta = \frac{1}{2}(\cos 2\theta + 1)$$

$$\sin^2\theta = \frac{1}{2}(1 - \cos 2\theta)$$

$$I = r^4 \int_0^{\frac{\pi}{2}} (\cos 2\theta + 1)(1 - \cos 2\theta) d\theta$$

$$= \frac{r^4}{4} \int_0^{\frac{\pi}{2}} (1 - \cos^2 2\theta) \, d\theta$$

$$= r^4 \int_0^{\frac{\pi}{2}} 1 \, d\theta - \frac{r^4}{2} \int_0^{\frac{\pi}{2}} (\cos 4\theta + 1) \, d\theta$$

Alien Pi in the Sky?

$$r^4{\tfrac{\pi}{2}\brack 0}[\theta] - \frac{r^4}{2}{\tfrac{\pi}{2}\brack 0}[\frac{sin4\theta}{4} + \theta]$$

$$= \pi\frac{r^4}{4}$$

$$I_y = \pi\frac{r^4}{4}$$

To assess how close to failure a particular beam is we may find the maximum stress associated with its deformation. Thus for compression we may have a buckling load

max buckling load
$$= \frac{\pi^2 EI}{L^2}$$

max stress bending mode
$$\sigma_{max} = \frac{Mc_1}{I}$$

max stress torque mode
$$\tau_{max} = \frac{Tr}{J}$$

material	E N/m²	σ_t N/m²	σ_y N/m²	v	ρ kg/m³
iron	2.06E+11	3.00E+08	2.50E+08	0.34	7.87E+03
aluminium	7.10E+10	8.00E+07	5.00E+07	0.33	2.71E+03
alu alloy	7.50E+10	6.00E+08	5.50E+08	0.3	2.5-2.9E+3
bronze 90:10	9.6 - 12E10	2.60E+08	1.40E+08	0.34	8.8 E+3
copper	1.17E+08	1.50E+08	7.50E+07	0.35	8.93E+03
Nickel alloy	1.10E+11	3.00E+08	6.00E+07	0.36	8.5 E+3
mild steel	2.10E+11	4.60E+08	3.00E+08	0.29	7.86E+03
stainless steel	2.05E+11	6.00E+08	2.30E+08	0.28	7.93E+03
Titanium	1.05E+11	4.13E+08	2.76E+08	0.3	4.54E+03

Table 2.1 Properties of metals

With these ideas in mind, I was able to develop a program to calculate expected displacements. As with all structural analysis it is vitally important to know in load bearing structures where the stresses lie and what the loaded configuration of the structure will be. This is important in understanding what is acceptable movement in the structure and for design of key components. Thus a bridge made of pliable material may be strong enough to carry any expected load but if during its use the movement is great enough to cause doubt on it's ability then users may lack confidence in its use. Another point demonstrating the need to know accurately the deformations under load would be when the load bearing structure is just part of a close fitting more complex structure possibly of composite materials. Thus deformations in the load bearer should be limited to what

Alien Pi in the Sky?

---

the various materials the composite structure will allow. Thus deformations should be well within the elastic limit of the material of a member and all stresses within it should be much lower than its yield stress σ_y

The code of this program is found as 3DENet.cpp in the subdirectory AlienPi\3DENet, examples of input and output files and a study of a series of increasingly complex configurations are available from this location.

Let us look at the program and how it works. The first thing to do is to draw our structure in terms of beams nodes and loads. For example take the simple lever illustrated below in figure 2.5, it comprises a short vertical stake fixed to the ground and a 0.5 metre arm that has a 10newton load applied vertically at the end. If we assume we use steel rod radius 5mm with E=2.05 e11, I=4.91 e-10, A=7.85 e-5, J=9.82 e-10, v=0.29.

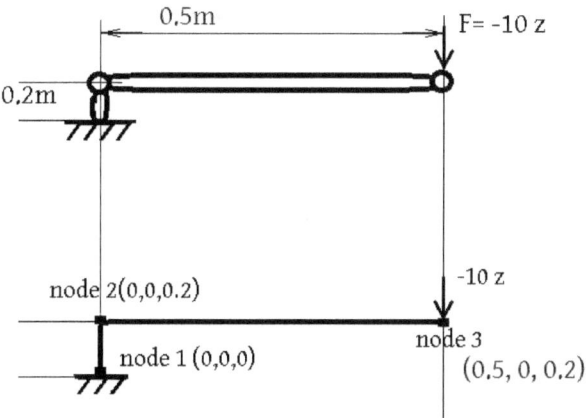

Figure 2.5 simple lever

It is easy to see the load of 10 newtons will cause bending in the two beams of the structure with a moment M=10*0.5 Nm. The deflection in a beam is given

$$\delta l = \frac{Ml^2}{2EI}$$

This will give a deflection of beam 1
δl=10*0.5*(0.2)²/(2*2.05e11*4.91e-10)

=9.93 e-4

In beam 2 δl= 6.21e-3

Alien Pi in the Sky?

From the diagram it is obvious that deflection of node 2 in beam 1 is in the positive x direction and deflection of node 3 in beam 2 will be in – z direction.

To use our program we have to create an appropriate input file and name it 3DENetIn.txt so our program can pick it up for its inputs.

```
3DENetIn.txt - Notepad
File Edit Format View Help
no beams
Beam no snd end  E    I    A   J   v
nnodes  Nodes[i].nn Nodes[i].Pos dfx nsp  beam[12] fflag[] F[]

2
1       1       2       2.05e11 4.91e-10 7.85e-5 9.82e-10 0.29
2       2       3       2.05e11 4.91e-10 7.85e-5 9.82e-10 0.29

3
1
0.0     0.0     0.0
1       1
1       1       1
0.0     0.0     0.0

2
0.0     0.0     0.2
2       1       2
0       1       1
0.0     0.0     0.0

3
0.5     0.0     0.2
1       2
0       1       1
0.0     0.0     -10.0

1
1
2       1       2
0
```

Figure 2.6 Input file

Fig 2.6 shows the input file. It starts with a five line header. The first figure to be read is the integer number of beams in the structure. After is listed the beams with their properties.

The first integer is the beam no. The unique number assigned to the beam. The second integer is the start node and the third integer is the finish node. The next five rational inputs are properties of the beam in order of E, I, A cross section area, J the polar moment of inertia

$$for\ circular\ cross\ section\quad J = \frac{\pi R^4}{2}$$

The last input describing the beam properties, is Poisson's ratio v, used in relating bulk modulus G with Young's modulus E.

$$G = \frac{E}{2(1+v)}$$

After details of the two beams we have an integer, the number of nodes in the structure.

We then list nodes with their individual properties.

First we have the unique identifying number of the node.

The next line are bit flags taking values 0 or 1 describing how beams are connected in the node.

The first bit describes if the node is fixed in position val=1 or motile val=0 this bit has to be set if the node is foot of the structure, the second bit value describes if the connect beams rigidly allowing beams to flex and sustain bending val=1 or allowing the connected beams to hinge and angle freely val=0. The third and final flag describes if the node allows free rotation of the beams preventing any accumulation of torsion val=0 or if the beams are held rigid in rotation, transmitting the torsion forces through the structure. The next line of node input describes the vector force applied to the node in terms of Cartesian.

This is repeated for the 3 nodes of the structure.

Alien Pi in the Sky?

There is now a series of integers the first states the number of loaded nodes, followed by the number of feet in the structure, then for each loaded node there has to be a defined route from each foot to the load.

In our structure we have 1 foot and 1 load. The route is then defined by the series of beams linking the foot to the loaded node. The initial integer of the line states the number of beams in the series, then a list of their unique numbers, but this is directional in that if the direction opposes the natural way the beam has been described the beams number is recorded negative.

The next is description of what are called free ends. These are limbs of the structure that do not contribute to loading because all the nodes in them are unloaded. In our present structure we have none of these so the input is 0.

Creating the input file in the Aliens/3DENet folder and double clicking the file 3DEnet.exe produces the output file 3DENetOut.txt a copy of which is seen below.

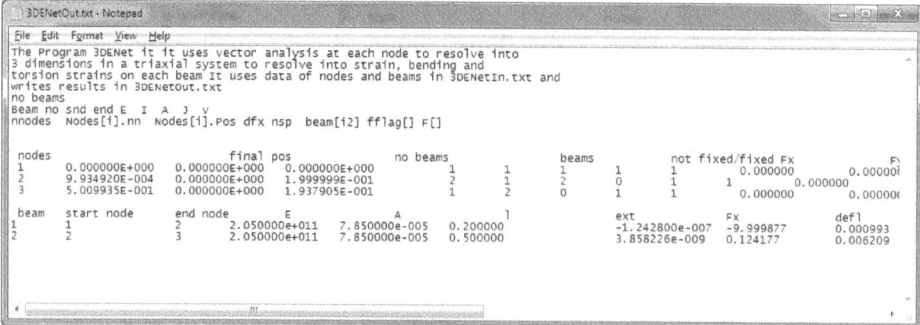

Fig 2.7 3DENetOut.txt

It confirms our free hand calculations showing a displacement of node 2 in the positive x direction, the deflection in beam 1 and the depression of node 3 in z direction with the deflection of beam 2. In addition it shows a small depression of node 2 in the z direction because of compressive force on beam 1. These values are all turned negative when we reverse the direction of force by changing it to +10N. Try changing the load to -10y. Thus we expect beam 1 to be in torsion and beam 2 to be in bending. The torsion in beam 1 is given

$$\frac{d\theta}{dl} = \frac{T}{GJ} = -0.0641$$

This agrees with the value theta/l obtained in the output file. And so by changing the direction of the load we are able to demonstrate the three modes of deformation in this simple structure. In looking at the stability of the structure we can use the three modes of failure and see which is appropriate to be examined according to the loading. Thus an X directional force will cause bending in beam 1 and compression or tension in beam 2. A Y directional force causing torsion in beam 1 and bending in beam 2. The limits are to be well within

$$\frac{\pi^2 EI}{L^2} = 3.97E+03 \text{ for compression,}$$

or σ_t for tension< 4.6e+8.

max stress bending occurs with Y or Z directed forces

$$\sigma_{max} = \frac{Mr}{I} = 5.09e7$$

Alien Pi in the Sky?

---

max stress torque occurs in beam 1 with Y directed force

$$\tau_{max} = \frac{Tr}{J} = 2.54e7$$

These are all well within the yield stress for steel, $\sigma_y = 3.0e8$

Let us look into a slightly more sophisticated example created from 5 lengths connected end on end. The copper alloy they are created from is such that it has E 1.25e+11 and circular cross section of radius 7.5mm. The first beam connects to foot vertically and the cantilever proceeds in +ve x direction as shown in fig 2.8

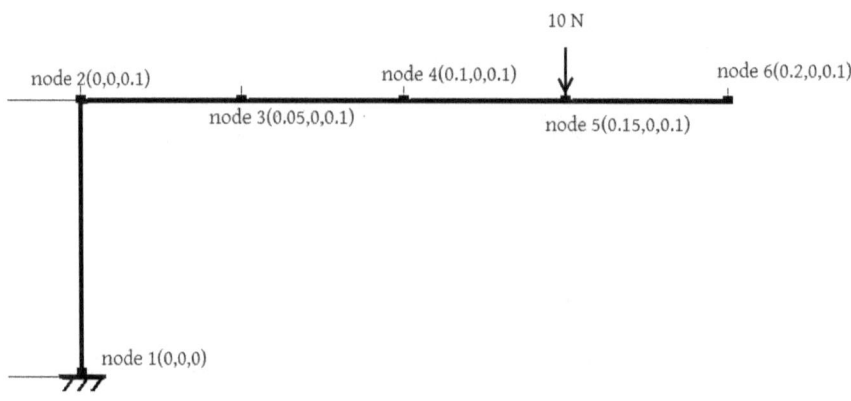

Fig 2.8 5 Beam cantilever

30

It is noted that the load is applied on node 5 of the structure, making beam 5 between node 5 and node 6 unloaded and what we call a free end. This is added at the end of the input file.

The input file to describe this set-up looks like

Beam no snd end E I A J v

nnodes Nodes[i].nn Nodes[i].Pos dfx nsp beam[i2] fflag[] F[]

5

1	1	2	1.25e11	2.48e-9	1.77e-4	4.97e-9	0.34
2	2	3	1.25e11	2.48e-9	1.77e-4	4.97e-9	0.34
3	3	4	1.25e11	2.48e-9	1.77e-4	4.97e-9	0.34
4	4	5	1.25e11	2.48e-9	1.77e-4	4.97e-9	0.34
5	5	6	1.25e11	2.48e-9	1.77e-4	4.97e-9	0.34

6

1

0.0   0.0   0.0

Alien Pi in the Sky?

1   1
1   1   1
0.0   0.0   0.0

2

0.0   0.0   0.10
2   1   2
0   1   1
0.0   0.0   0.0

3

0.05   0.0   0.10
2   2   3
0   1   1
0.0   0.0   0.0

4

0.10   0.0   0.10
2   3   4

0	1	1
0.0	0.0	0.0

5

0.15	0.0	0.10
2	4	5
0	1	1
0.0	0.0	-20.0

6

0.20	0.0	0.10
1	5	
0	1	1
0.0	0.0	0.0

1

1

4	1	2	3	4

Alien Pi in the Sky?

1

1   5

We may run the program with adjustments to the loads to illustrate as before what we may expect from loads directed in pure z, x, or y directions.

We may compare the result of the cantilever with that of when both ends of the structure are fixed. To convert the input file to reflect this situation node 6 will have flag 1 set 1, the number of feet is set to 2, a second path has to be indicated from the second foot to the load and the free end has been converted to a foot has to be reset to 0. In effect the change of the input file from node 6 onwards is replaced with the values shown below.

6

0.20   0.0   0.10

1   5

1   1   1

0.0   0.0   0.0

2

1

Olatunde Adeyemo

4	1	2	3	4
1	-5			

0

Comparing output files we find maximum moments for either configuration for a 20N load is of the magnitude of 3Nm. This results in $\sigma_{max} = \frac{Mr}{I}$ =9.07e6 and $\tau_{max} = \frac{Tr}{J}$ =4.54e6. These are less than σ_y=7.5e7 for copper and the structure will hold well. We can equally assess if the structure will suffer buckling or tensile failure by comparing the beam with greatest inline force, read as Fx in the output file and calculate the stress as Fx/A and comparing it with the buckling stress $\frac{\pi^2 EI}{L^2}$ = 2.23e5. L here is considered to be 0.15m since lengths of beams 2,3,4 are in line. With Fx= 20 N in –X direction Fx/A =1.13 E5 we expect the structure not to buckle but it is close to the limit and we may consider the safety factor to be insufficient.

It can equally be shown that the program is able to cope with multiple loads applied at different nodes. To elaborate suppose we add -10N in Z at node 3 and +10N in Y at node 4. The input file will change for node 3 onwards to

Alien Pi in the Sky?

3

0.05   0.0   0.10

2   2   3

0   1   1

0.0   0.0   -10.0

4

0.10   0.0   0.10

2   3   4

0   1   1

0.0   10.0   0.0

5

0.15   0.0   0.10

2   4   5

0   1   1

0.0   0.0   -20.0

6

Olatunde Adeyemo

0.20 0.0 0.10

1   5

1   1   1

0.0   0.0 0.0

2

3

2   1   2

3   -5   -4   -3

3   1   2   3

2   -5   -4

4   1   2   3   4

1   -5

0

A study of these developments showing the results of running the program with the different configurations described above are demonstrated in the file /AlienPi/3DNet/3DNetOutstudy.txt which is a compilation

Alien Pi in the Sky?

of output files as created by the program. To demonstrate how the program can deal with a sophisticated 3d, 2 footed and multiple loaded structure as in fig 2.9. The input file representing the structure was created and the program was run to get results.

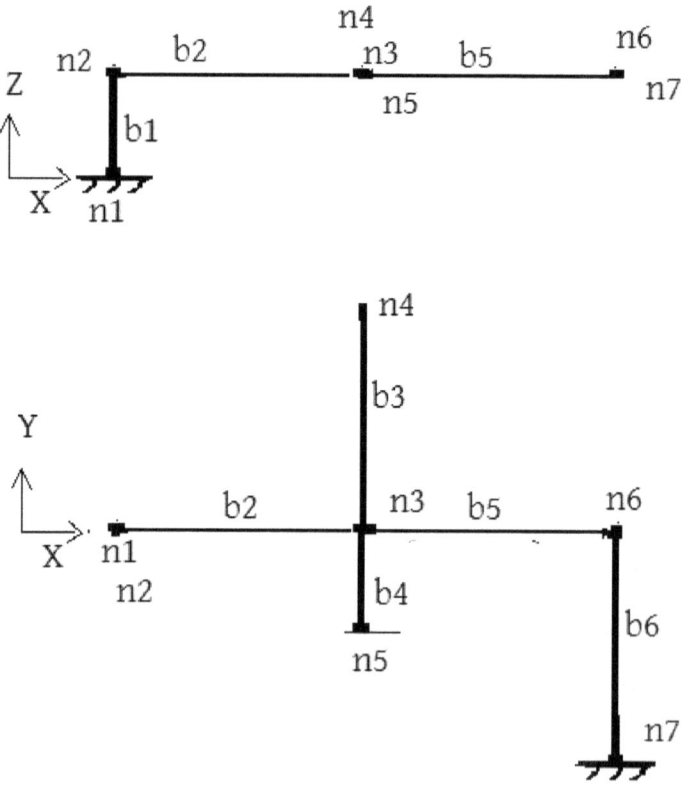

Fig 2.9 A complex multi loaded structure

In this structure the loads are applied on nodes 4, 3 and 5 and the feet are nodes 1 and 7.

The original position of nodes are shown in the following table.

node	x	y	z
1	0	0	0
2	0	0	0.2
3	0.5	0	0.2
4	0.5	0.5	0.2
5	0.5	-0.2	0.2
6	1.0	0.0	0.2
7	1.0	-0.5	0.2

Assumptions are made on properties of the beams such that they all have E=2.05e11, I=2.48e-9, J=4.97e-9 and v=0.34.

With nodes 3,4 and 5 loaded and set as below

Alien Pi in the Sky?

3

0.5 0.0 0.2

3 2 3 5

0 1 1

0.0 0.0 -20.0

4

0.5 0.5 0.2

1 4

0 1 1

-10.0 0.0 10.0

5

0.5 -0.2 0.2

1 4

0 1 1

0.0 0.0 -20.0

The full input file is shown in the file complexin.txt results are shown in the file /Alien/3DENet/complexout.txt.

Olatunde Adeyemo

This can be broken down to analysing this as a bearing structure of beams 1, 2, 5 and 6 anchored at both ends with loads loaded on cross member made from beams 3 and 4 that carry loads. It is obvious from examining the structure that beams 1,2,5 and 6 will bear loads through bending and possibly beams 2 and 5 may suffer torsion. Beams 2 and 4 will suffer bending moments directly from the arrangement of loads they carry. All this is observed in the detail of the solution file.

Alien Pi in the Sky?

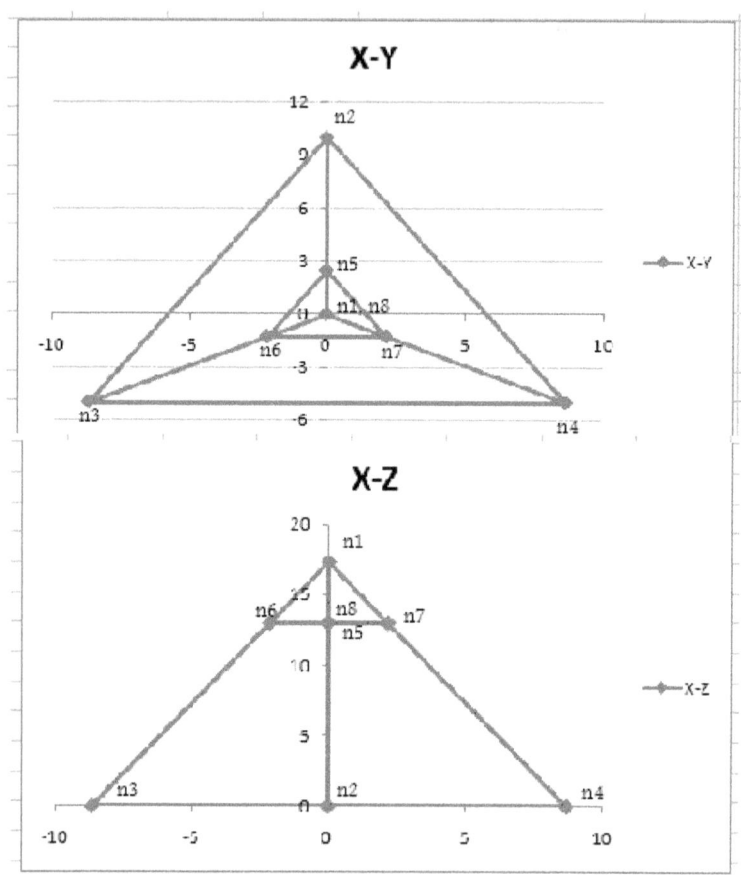

Figure 2.10 Tripod Frame

Finally to show how the program copes with 3d, I set up a structure similar to tripod with an equilateral base. It represents a large structure 17.3m tall, members are of radius 25mm steel with E=1.25e11. A 30N load centrally loaded on it will stress it but still within our assumptions of small deviations. The input and output files that show this are tripodin.txt and tripodout.txt. We may like to see which members are most prone to failure and what load would cause this.

We may observe the highest compression stresses are found in beams 4, 8 and 9 each having length of 15m. For a 15m beam buckling stress $\frac{\pi^2 EI}{L^2}$ = 1683Nm^{-2} which is much less than the indicated stress of 4469N m^{-2} and would indicate failure. It might be more practical to limit the length of beams 4,8 and 9 to half their present length to get them within parameters of avoiding buckling failure. The greatest deflections are found in beams 4, 8 and 9 with a maximum stress $\sigma_{max} = \frac{Mr}{I}$ =8.32e6 which compares with a σ_y=3.0e8 for steel. This would suggest to us that the structure is well will not fail through bending stress as presently loaded.

The fact of buckling failure in beams 4, 8 and 9 would require a redesign and a new evaluation in accordance with the changes made.

This program has other capabilities, that have not been demonstrated here, such as when a node in a limb allows angular movement between beams or allow swiveling to stop any torques building.

To conclude we have seen how we can use this program to look at complex loaded structures and test for movement and failure in its various limbs and allows the engineer to cheaply try alternative options and new designs towards his structural purpose.

Alien Pi in the Sky?

Chapter 3. Pi in the Sky and Extended Precision Calculation

My objective in this chapter is to develop an arithmetic system that can be used to calculate elaborate series equations to a desired precision described in terms of decimal places. This system will then be put to test by using it to calculate the value of pi, the ratio of the circumference of a circle to it's radius to a prescribed precision.

The first question is why do we have to derive a new system? In terms of everyday calculations we may use computers or calculators to perform the arithmetic. Depending on their storage per variable the precision is limited.eg in C++ a normal float variable of 2 bytes may comprise of a sign, an exponent of 8 bits making up ±E37 and a 22 bit mantissa (maximum value 4.2E6 i.e. correct to 6 decimal places (dp)). Equally a double may have an 8 bit exponent ±E37 and 52 bit mantissa(maximum value 4.5E15 correct to 15 or 16 dp). The problem with an arbitrary length mantissa in normal programming is that the mantissa beyond about 6 significant figures in a single precision float and beyond 15 significant figures in a double precision float is inaccurate and if used in calculations contributes nothing to the accuracy of the result. We need to be able to extend the precision indefinitely into the extended significant figures.

In our system of arithmetic we must first define our number. The real rational number we are to use is made up of a sign, a mantissa (which carries the precision) and an ordinate (the power of 10 to which the mantissa must be

Alien Pi in the Sky?

raised to make it correct). Ideally the mantissa has a value equal or greater than 1 but less than 10. Thus 1.0E15 has sign +1

mantissa 1.0

ordinate +15

The rest mass of an electron 9.10938291 E-31kg would be written with

sign +1

mantissa 9.10938291

ordinate -31

The sign and ordinates can be assigned as integers. To simplify computing purposes we can write the mantissa as an integer array. The benefits from this include saving of memory storage since each integer is just 16 bits as compared to 32 bits for a float with a corresponding simplification in arithmetic operations. The precision we require is held in the length of the mantissa Each integer represents 6 significant digits of the mantissa, 6 chosen for aesthetic reasons and manageability of the number, thus the full number of significant figures for representation is 6*ncells, where ncells is the number of dimensions of the mantissa array.

Having fabricated our real number as a composite of integers we can declare it as a structure NxlongR comprising an integer array mantissa, an integer exponent and an integer sign. This defines our n decimal place (dp) real number object NxlongR.

Olatunde Adeyemo

NxlongR takes some special values. When the sign takes the value 0 it indicates the NxlongR object has been evaluated to 0. When the exponent=12345 it indicates NxlongR object has been evaluated to division by 0, indeterminately large.

Now in real terms we demonstrate how useful this is.

Given the charge radius of a proton 0.8751×10^{-15}m then volume of a proton is $4/3\pi r^3 = 0.893 \times 10^{-45} m^3$

The size of the observable universe radius 4.65×10^{10} light years $= 4.65e10 \ x2.99e8 \ x3.6e3 \ x24 \ x365 = 4.4 \times 10^5 \ x10^{21} = 4.4 \times 10^{26}$m.

The ratio of volume of a proton to the known universe

$$= \left(\frac{4.4 * 10^{26}}{8.8 * 10^{-16}}\right)^3 = 1.3 \times 10^{125}$$

In other words 126 decimal places would be enough to describe the volume of the known universe down to the accuracy of the volume of a proton. Clearly this accuracy is the best that we could ever be expected to practically use. However as a thought experiment numbers like e, π or surds like

$\sqrt{2}, \sqrt{3}, \sqrt{5}, \sqrt[3]{3}$ etc can be generated to whatever accuracy we require. In this chapter I shall show how this can be performed.

Alien Pi in the Sky?

As an aside, if we look at our progress over the past hundred years concerning space exploration we can consider we are making the first halting steps to be unconstrained by our planet. What would be the most desirable gift to mankind apart from eternal life? Why of course it would be inexhaustible sustainable energy. The thing is we have a source of this in sunlight. The problem at present is we have not developed enough to capture, store and utilise it as it is. In another 1000 years we come back and look at ourselves, each of us is using untold gigajoules of energy to transport ourselves to far flung corners of the solar system and maintain habitable environments in the most extreme, hostile and remote locations. Where are we getting all this energy? Why, we found a way of converting a good portion of the solar energy, storing it and transporting it to where it is to be used. The thing is, an alien from 100 light years away would have noticed that over the period our sun has disappeared or gone dim for an unknown reason. If they were unable to witness the decline in radiance, they might well say that the solar system posed a question of part of the missing mass of the universe, knowing from the motion of neighbouring stars that something must be there, but the nature of the brightness and spectrum received confusing their own star models.

It is interesting in that we know from observation of galaxy and star motion that about 90% of the mass of the universe is unobserved. Could some of this be due to shading caused by alien methods of solar harvesting? Of course if this were the case we would have to take a completely different approach to detecting these aliens.

To return to our extended precision calculation we must now create basic arithmetical and manipulative functions and routines for handling and maintaining the NxlongR real number structure.

Rippleup   void ripppleup(a* NxlongR)

Is a function that takes a pointer to a NxlongR structure that is necessary to eliminate any array element of the mantissa that becomes greater than 999999 in any manipulation. Thus for example, if after manipulation mantissa array element x is 21 000000 this is not acceptable and has to be converted to 000000 with a carry over of 21 to be added to array element x-1.

xnegc   void xnegc(NxlongR* c,int dp)

Is a function to ensure all mantissa elements are converted to positive. If for example after an manipulation element x becomes -234124, xnegc will cause array element to become 765877 and array element x-1 drops by 1. In this function are measures to ensure that the first element of the mantissa lies between 100000 and 999999.

addXlr   void addXlr(NxlongR a, NxlongR b,NxlongR* c,int dp)

Is the function that adds two NxlongR structures. This function has to determine which of the two numbers is greater in absolute terms, choose the greater one and align the smaller against it before adding the two numbers

Alien Pi in the Sky?

together. Signs must be taken into account. It uses rippleup and xnegc to mop and clean up results.

It is noted that there is no function equivalent to subtraction. It is not necessary since to perform this we simply reverse the sign of the second number and add.

prodxlr   void prodxlr(NxlongR a, NxlongR b,NxlongR* c,int dp)

Is the function that multiplies two NxlongR objects. This has a way of detecting whether one of the multiplicands has previously been evaluated to zero and thus transferring this to the product by evaluating the sign of that object sgn=0 .

The mechanism of multiplication is performed by taking a matrix element from each of the multiplicands and knowing they can range from 999999 and 000000.

$$999999 = 10^6-1$$

the max of multiplication is $(10^6-1)*(10^6-1) = 10^{12}-2 \times 10^6 + 1 = 999998000001$

the lowest non zero value is 000001.

The value of the product the two elements is to be calculated and aligned onto the answer NxlongR object.

Divlxr   void Divlxr(NxlongR divisor,NxlongR divend,NxlongR* quot,int dp)

Is a function using classic long division to divide a NxlongR dividend, by a NxlongR divisor. Long division relies on making an estimate for the answer, using this estimate to multiply the divisor and comparing the result by subtracting it from the dividend. The result is then saved. The next estimate is formed when an evaluation of the remainder divided by the divisor is added to the solution.

By following through this method using NxlongR objects for dividend, divisor and quotient we are able to build an accurate quotient at 12 decimal places per cycle. If for some reason the divisor evaluates to zero the quotient is evaluated to "out of scope" by setting exp = 12345 .

itoxR   void itoxR(int intg,NxlongR* xlo)

This function reads an integer and converts it into an NxlongR object. This function enables a quick conversion from normal integers into a NxlongR object we are able to manipulate with the extended precision system.

invtan   void invtan(NxlongR x,NxlongR* Itan,int dp)

The series

$$\theta = x - \frac{x^3}{3} + \frac{x^5}{5} - \frac{x^7}{7} ...$$

Is the Gregory series that converges to $\tan^{-1}(x)$ if $-1 \le x \le 1$.

We know in radial measurement $\tan^{-1}(1) = \pi/4$ however convergence of $\tan^{-1}$ series for $|x| > 0.5$ is very slow and impractical, when later we want to get a value for π it cannot be used effectively directly. This series is still very

Alien Pi in the Sky?

important in evaluation of π as will be explained. As such a special function invtan that produces the result of this series as a NxlongR object is still very useful to our project.

crstrXlr   void crstrXlr(NxlongR xlr,int sf,char* ostr)

Is a routine used for output of a NxlongR object. It produces a string ready for printing to a console or a file. For ease of reading the mantissa is written in lines 60 dp wide in groups of 6 characters representing the content of a corresponding array element.

NRoot   void NRoot(NxlongR A, int N,NxlongR* x,int Ndp)

NRoot is the extended precision routine that iterates down on the real positive solution to $(x^n - a) = 0$ where a is a positive value. This uses the relation $f(x) = (x^n - a) = 0$.

$$f(x) = x^n - a = 0$$

$$f'(x) = n\, x^{n-1}$$

$$f(x + \varepsilon) \cong f(x) + \varepsilon f'(x)$$

$$\cong 0 + \varepsilon n\, x^{n-1}$$

$$\text{if } \varepsilon = \epsilon x, \text{where } \epsilon \ll 1$$

$$f\big(x(1 + \epsilon)\big) \cong \epsilon\, n\, x^n = \epsilon n a$$

$$\text{if for estimated soln } x_1$$

$$f(x_1) = y \cong \epsilon n a$$

$$\epsilon \cong \frac{y}{na}$$

$$x_1 \cong x(1+\epsilon)$$

$$x \cong \frac{x_1}{1+\epsilon} \cong x_1(1-\epsilon)$$

$$a\ better\ solution\ is\ thus\ x_2 = x_1(1-\frac{y}{na})$$

We can use x_2 as a new estimate for x and continue the iteration cycle.

The beauty of this solution is that it produces our resultant surd answer ($\sqrt{3}$, $\sqrt{7}$, $^3\sqrt{3}$ etc) using only the "easy" arithmetic of multiplication and the new solution is based on the error of the function and the estimate is procured with reduced operations.

This solution may appear to be very powerful, however for it to converge there are conditions on the initial estimate. Generally if the magnitude of a is greater than 0.33 an initial estimate of x=1 will converge, values of a less than 0.80 use an estimate of a/n to converge and when the magnitude of a is less than 1e-4 we get a closer estimate by setting the exponent x to (exponent of a)/N.

Using the core program and functions we may perform a study to evaluate surds to demonstrate how well this system copes with operations. To this object the code to be found in the programs rtNA.cpp and Roots.cpp were produced. These can be located under the subdirectory branch of my website \AlienPi\PI\Root .

Alien Pi in the Sky?

Roots finds the roots of 0.5,1.5,2...10.5 incremented by 0.5 nominally to 400 dp. The executable and output files for this program are found in the subdirectory ...\PI\Root\Roots . The ouput is named Roots.txt.

rtNA is interactive and demands the real positive number you wish to get the root of and produces a root accurate to 600dp. The executable and output files for this program are found in the subdirectory ...\PI\Root\RootNA . The ouput is named Root.txt. In the same directory a whole series of results have been manually assembled into a file to show the consistency of answers in the file surdsW.txt .

Study of Surds

To test our extended precision system we can use it to generate surds to a required precision. We can easily test the result by multiplying through the obtained solution and looking at the accuracy of the answer.

Surds by their nature often produce completely random sequences of digits that can be fundamental in generating random numbers and can be used in cryptography and coding.

The results of repeated calculations of this type for root 2 and root 3 are available for examination in the file surdsW.txt found in directory .\AlienPi/PI/Roots it shows that although we have a nominal accuracy of 400 dp, because inbuilt safety and provision for loss the answers are consistently correct to between 432 and 436 dp proving the effectiveness and accuracy of the routines involved.

Evaluation of Pi-1

Having established an arithmetic system capable of calculating to whatever precision required we can now apply it to our main objective which is to calculate Pi (π) to extended accuracy.

In our first method we attack it directly using geometry as a visual aid to build closer and closer approximations. We know π as the ratio of the circumference of a circle to its diameter, thus we know the basic equation circumference =2 πr, r radius of circle, diameter 2*r.

To start we use values of bounding internal and external squares to create an estimate.

Alien Pi in the Sky?

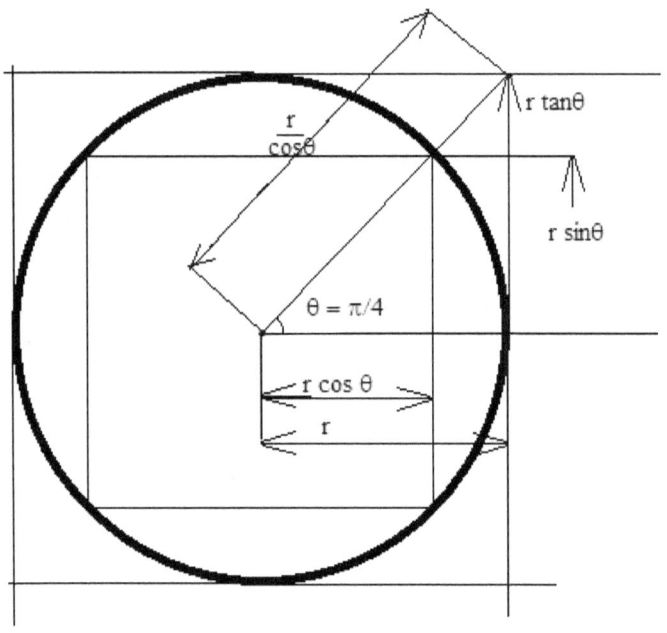

Fig 3.1 Bounding internal and external squares

From fig 3.1 it is easily seen that the sides of the external bounding square sum to 8r*tan(π/4) and for the bounding internal square is 8r*sin(π/4) , the average between them is then 4r*tan(π/4)+4r*sin(π/4). Knowing tan(π/4) = 1 and sin (π/4) = 1/√2 these values are calculated, giving an estimate for π =2*(1+1/√2) =2+ √2 = 3.4142 , accurate to a single dp.

Olatunde Adeyemo

If we halve the angle we double the number of sides and get a closer approximation. We can produce relations for sin(θ/2) and tan(θ/2) as long as these are accurately calculated we should get an improvement in our estimate, repeating this process multiple times we can produce closer and closer estimates.

For sin(θ/2) we use relation

$$\cos(\theta \pm \phi) = \cos\theta\cos\phi \mp \sin\theta\sin\phi$$

$$\cos(\theta - \phi) - \cos(\theta + \phi) = 2\sin\theta\sin\phi$$

If $\phi = \theta$

$$1 - \cos(2\theta) = 2\sin^2\theta$$

Replace 2θ with θ

$$1 - \cos(\theta) = 2\sin^2\frac{\theta}{2}$$

$$\sin\frac{\theta}{2} = \sqrt{\frac{1 - \sqrt{1 - \sin^2\theta}}{2}}$$

For tan(θ/2) we use relations

Alien Pi in the Sky?

$$\sin(\theta \pm \phi) = \sin\theta\cos\phi \pm \sin\phi\cos\theta$$

$$\cos(\theta \pm \phi) = \cos\theta\cos\phi \mp \sin\theta\sin\phi$$

$$tan(\theta + \phi) = \frac{\sin\theta\cos\phi + \sin\phi\cos\theta}{\cos\theta\cos\phi - \sin\theta\sin\phi}$$

$$tan(\theta + \phi) = \frac{\tan\theta + \tan\phi}{1 - \tan\theta\tan\phi}$$

If $\phi = \theta$

$$tan(2\theta) = \frac{2\tan\theta}{1 - \tan^2\theta}$$

Replace 2θ with θ

$$tan(\theta) = \frac{2\tan\frac{\theta}{2}}{1 - \tan^2\frac{\theta}{2}}$$

$$tan(\theta)\left(1 - \tan^2\frac{\theta}{2}\right) - 2\tan\frac{\theta}{2} = 0$$

quadratic in $\tan\frac{\theta}{2}$

$$\tan\frac{\theta}{2} = \frac{1 \pm \sqrt{1 + \tan^2\theta}}{\tan\theta}$$

$$\tan\frac{\theta}{2} = -\frac{1}{\tan\theta}(1 \pm \sec\theta)$$

$$\tan\frac{\theta}{2} = \frac{1}{\sin\theta} - \frac{1}{\tan\theta}$$

$$\sin\frac{\theta}{2} = \frac{\tan\frac{\theta}{2}}{\sqrt{1 + \tan^2\frac{\theta}{2}}}$$

From this analysis we must first be able to calculate $\sqrt{2}$ to the accuracy in significant figures required and we need to be able to calculate 2nd roots to the number of dp required, It is also obvious that the root of a root required to calculate $\sin\theta/2$ from $\sin\theta$ is quite complicated and will stress the accuracy of our solution. We can get around this by calculating $\sin\theta/2$ from $\tan\theta/2$ after it has been created. For this reason in calculating $\tan\theta/2$ we derive it directly from $\tan\theta$ and avoid using $\sin\theta$ or $\sin\theta/2$.

To get this to work we can show our working in a flow chart.

Alien Pi in the Sky?

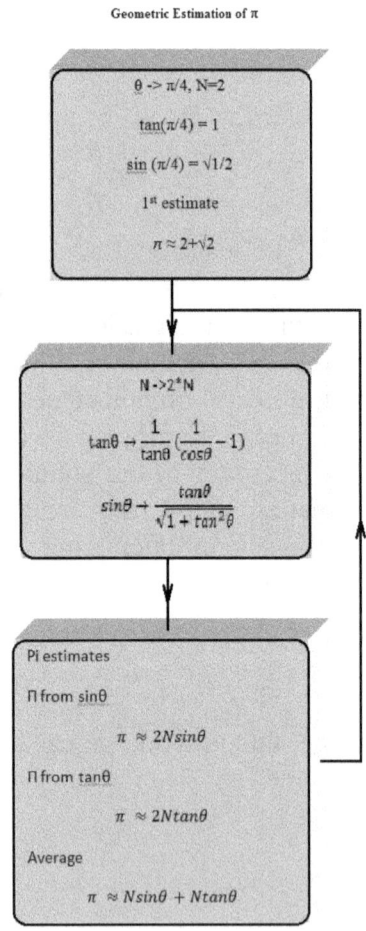

Figure 3.2 Geometric Estimation of Pi

Olatunde Adeyemo

Essentially in each cycle we are doubling the number of sides of our bounding polygon.

We see how this works with the program \Alien\GeoII\main.cpp which uses the core functions described. It does the iteration 60 times and writes the answer in \Alien\GeoII\GeoPi.txt.

An examination of GeoPi.txt shows the average of the estimates of sin and tan gives the best estimate being normally better by 1 dp than each of the other estimates. After 60 iterations we are able to get Pi correct to 37dp from a polygon of 2 305843 009213 693952 sides. This is clearly a slow convergence, but the exercise demonstrates the principle.

One way of improving the estimate is to use convergence to the solution. Knowing we are seeking the solution tanθ -1=0

if tanθ =1

$$f = 0 = tan\theta - 1$$
$$f' = sec^2\theta = 1 + tan^2\theta$$

From Newton Raphson expansion

Alien Pi in the Sky?

$$f(\theta + \varepsilon) \approx f(\theta) + \varepsilon f'(\theta)$$

then

$$y = \tan(\theta + \varepsilon) - 1 \approx \tan\theta - 1 + \varepsilon(1 + \tan^2\theta)$$

$$\varepsilon \approx \frac{y + 1 - \tan\theta}{1 + \tan^2\theta}$$

This means

Given an estimated solution x_1 giving f=y

A better solution will be obtained by

$$x_2 = x_1 - \frac{y + 1 - \tan\theta}{1 + \tan^2\theta}$$

Since $\tan\theta = 1$

$$x_2 = x_1 - \frac{y}{2}$$

Putting this together in a program we can use the geometrical approximation to get close to the solution and subsequently use the convergence formula to get ever closer approaches to the solution. This has been performed in the program GeoPi2.cpp and the result of run can be seen in the solution file GeoTanPi.txt these are to be found in the directory \AlienPi\PI\GeoPi. The program uses 15 iterations of the geometrical approximation down to an angle of $\pi/131072$, giving an answer for π accurate to 10 significant figures. The program then goes on to make iterative corrections to the solution of $\tan\theta = 1$. In each iteration the value of the calculated values of $\sin\theta$ and $\cos\theta$ are printed into the output file to demonstrate their convergence to $1/\sqrt{2}$, it is seen that 9 iterations lead to $\sin\theta$ and $\cos\theta$

differing in the 520 dp and the 10th they are both correct to all 600dp the solution for π is correct to 592 dp and takes about 11m 46s on a dated 32bit dual core 1.5Ghz machine.

The main drawback of this method is the accurate calculation of tanθ. To achieve this we have the series generation of sinθ and cosθ.

$$cos\theta = 1 - \frac{\theta^2}{2!} + \frac{\theta^4}{4!} - \frac{\theta^6}{6!} ... + (-1)^{\frac{2n}{2}} \frac{\theta^{2n}}{2n!}$$

and

$$sin\theta = \theta - \frac{\theta^3}{3!} + \frac{\theta^5}{5!} - \frac{\theta^7}{7!} ... + (-1)^{\frac{2n}{2}} \frac{\theta^{2n+1}}{(2n+1)!}$$

This is performed by the introduced function CosSin

CosSin. void CosSin(NxlongR x,int cs,int Ndp,NxlongR* CoS)

These series each have to be generated for each iteration of the convergence method, this is very demanding on computing as the required number of significant figures increases. It is effectively a multiplication of the work to be done.

Alien Pi in the Sky?

Evaluation of Pi-2

The previous section shows a visual and tangible method for the estimation of π. We find it has a number of deficiencies. Firstly even when evaluating $\tan\theta/2$ we are serially applying roots that are irrational functions requiring iterative solutions and cumulative errors. Secondly the convergence is linear, slow and inadequate for a high power solution. When applying the convergence formula we need to do a full expansion for $\sin\theta$ and $\cos\theta$ for each cycle. In the evaluation we do here we seek a mathematical solution based on the expansion of $\tan^{-1}(1) = \pi/4$, we have already shown the Gregory series for the expansion of $\tan^{-1}(x)$ by function invtan. It is easily seen that applying the expansion directly on the value 1.0 would be ineffectual because it is close to the boundary of convergence.

Consider the following analysis.

from $\tan(\theta+\phi)$

$$\tan(2\theta) = \frac{2\tan\theta}{1 - \tan^2\theta}$$

if $\tan\theta = \frac{1}{5}$

$$\tan 2\theta = \frac{\frac{2}{5}}{1 - \frac{1}{25}} = \frac{\frac{2}{5}}{\frac{24}{25}} = \frac{5}{12}$$

$$\tan 4\theta = \frac{\frac{10}{12}}{1 - \frac{25}{144}} = \frac{\frac{5}{6}}{\frac{119}{144}} = \frac{120}{119}$$

Remember

$$tan(\psi + \phi) = \frac{\tan\psi + \tan\phi}{1 - \tan\psi\tan\phi}$$

if $\tan \psi = 1$

$$tan(\psi + \phi) = \frac{1 + \tan\phi}{1 - \tan\phi} = \frac{120}{119} = \tan 4\theta$$

solve for $\tan\phi$

$$1 + \tan\phi = (1 - \tan\phi) * \frac{120}{119}$$

$$\tan\phi \left(1 + \frac{120}{119}\right) = \frac{1}{119}$$

$$\tan\phi = \frac{1}{119} * \frac{119}{239} = \frac{1}{239}$$

$$4\theta = \psi + \phi$$

$\psi = 4\theta - \phi = \pi/4$

Since $\theta = \tan^{-1}\frac{1}{5}$ and $\phi = \tan^{-1}\frac{1}{239}$

They are both easily obtained from the Gregory series.

Giving

$$\pi = 16 \tan^{-1}\left(\frac{1}{5}\right) - 4\tan^{-1}\left(\frac{1}{239}\right)$$

This formula was devised by Machin to generate π. Using the Machin formula we can use the Gregory series effectively as both parts form a series of rapidly reducing

Alien Pi in the Sky?

terms. The one consideration to be upheld is that each term has to be evaluated up to the last dp of required precision.

An implementation of this formula is made in the program corrtn1.cpp found in the directory \AlienPI\PI\corrctn1 with the result written to the output file corrctn1.txt. This program has nominally been set to calculate π to about 6000 places. This was performed on a 32bit dual core 2.8GHz machine that took about 12.5 hours to calculate π accurate to 5834 significant figures. Test runs and faster computations with less decimal places can be made by adjusting the value of variable Ndp found in the function main of this program. If the program is required to calculate more decimal places the constant ncells found in the header of the program should be increased and the character string variable "istring" found in function "main" would have to be increased for the full answer to fit. The program was devised in a "netbeans" C++ IDE using Netbeans 8.0.1 using it or revising or refining it in any other IDE may require alterations or refinement.

You are encouraged to use the functions and programs to your own for your own purpose with acknowledgement.

Olatunde Adeyemo

Bibliography

Cambridge University Engineering Department Materials Data Book 2003 edition

Arthur M.Wahl Mechanical springs, Cleveland, O., Penton Pub. Co., 1944. http://hdl.handle.net/2027/uc1.$b76475

Jerry Miller π to 500K Decimal Places Bandanna Books © 2010

Olatunde O Adeyemo A Little Rocket Science and other Baubles 2013 www.lulu.com

www.engineeringtoolbox.com/metals-piossns-ratio-d_1268.html

www.angelfire.com/my/welding/metal.html

Alien Pi in the Sky?

http://www.diracdelta.co.uk/science/source/y/o/youngs%20modulus/source.html#.Vs5EW32LRdg

Mechanical Properties of Carbon Fiber Composite Materials, Fiber / Epoxy resin (20°C Cure)

https%3A%2F%2Fwww.acpsales.com%2Fupload%2FMechanical-Properties-of-Carbon-Fiber-Composite-Materials.pdf&usg=AFQjCNEWhmCiz6fXBh4_fdmfNSDqEbzRGg&sig2=SflcA-FjlPdsPM-OAglTFQ&bvm=bv.127178174,d.ZGg

www.ingramcontent.com/pod-product-compliance
Lightning Source LLC
Chambersburg PA
CBHW072234170526
45158CB00002BA/889